E R U L E A

880

This plant was bought
at a sale at Stevens
of plants imported
by the New Plant
& Bulb Company
of Colchester
last June

KEW POCKETBOOKS

ORCHIDS

Curated by Michael F. Fay and Gina Fullerlove

Kew Publishing
Royal Botanic Gardens, Kew

KEW HOLDS ONE OF THE LARGEST COLLECTIONS of botanical literature, art and archive material in the world. The library comprises 185,000 monographs and rare books, around 150,000 pamphlets, 5,000 serial titles and 25,000 maps. The Archives contain vast collections relating to Kew's long history as a global centre of plant information and a nationally important botanic garden including 7 million letters, lists, field notebooks, diaries and manuscript pages.

The Illustrations Collection comprises 200,000 watercolours, oils, prints and drawings, assembled over the last 200 years, forming an exceptional visual record of plants and fungi. Works include those of the great masters of botanical illustration such as Ehret, Redouté, the Bauer brothers, Thomas Duncanson, George Bond and Walter Hood Fitch. Our special collections include historic and contemporary originals prepared for *Curtis's Botanical Magazine*, the work of Margaret Meen, Thomas Baines, Margaret Mee, Joseph Hooker's Indian sketches, Edouard Morren's bromeliad paintings, 'Company School' works commissioned from Indian artists by Roxburgh, Wallich, Royle and others, and the Marianne North Collection, housed in the gallery named after her in Kew Gardens.

INTRODUCTION

WITH AN ESTIMATED 28,000 SPECIES
worldwide, orchids (Orchidaceae) are one of the
largest plant families, rivalled only by the daisy family
(Asteraceae). Orchids occur on all major landmasses,
except Antarctica, and also on many islands. They grow
in a wide range of habitats from tundra to tropical
rainforests and from sea level to over 4,000 m (13,123 ft),
only being absent from true deserts. They also have
remarkably diverse lifestyles – some occur in soil, but
others grow on trees or rocks. They are divided into five
subfamilies – the smallest, Apostasioideae (page 12)
containing fewer than 20 species in two genera.

The amazing diversity and strange life histories of
orchids have made them the subject of fascination for
many scientists and horticulturists. Charles Darwin,
for example, wrote to his friend Sir Joseph Hooker
(the second Director of Kew) that 'I never was more
interested in my life in any subject than this of orchids'.
This fascination has continued at Kew and other
institutions since the time of Darwin and Hooker –
scientists study their taxonomy, relationships and
interactions with other organisms, and horticulturists

develop cultivation methods for an increasing range of species. More recently, these studies have produced results subsequently used in conservation programmes, for example *Cypripedium calceolus* (page 16).

In nature, orchids rely on a relationship with fungi that provides the nutrients necessary for seed germination and the early stages of their development. Many continue this affiliation for their whole life, balancing food from the fungus with what they produce by photosynthesis; in extreme cases some, such as *Epipogium aphyllum* (page 47), have even lost the ability to produce chlorophyll and therefore do not photosynthesise.

Nearly all orchids are dependent for pollination on animals, which include flies, bees, butterflies, moths, e.g. *Angraecum sesquipedale* (page 91), birds, e.g. *Cattleya cernua* and *Masdevallia rosea* (pages 72 and 79), bats and honey possums (marsupials). As a result of this wide range of pollinators, the flowers range from delicate structures only a few millimetres across to large flowers tough enough to surviving probing beaks of birds or tongues of bats and marsupials. Some orchids provide a reward to their pollinators, but many orchids attract them by appearing to offer a reward ('pollination by deceit'). In species that do not provide a food reward, the flowers often resemble those of unrelated plants that do. In addition to food deception, many other pollination syndromes exist, and some species even practise 'sexual deception' with the flowers resembling and smelling like female insects, attracting males that attempt to copulate with them and pollinating the flowers in the process e.g. *Ophrys speculum* (page 39).

From the 18th century, the cultivation of tropical orchid species became increasingly popular, but it was largely a hobby for wealthy people able to afford construction and heating of greenhouses. They often also commissioned plant collectors to find plants for their collections. *Calanthe* [or *Phaius*] *tankervilleae* was probably the first tropical orchid species to flower in the UK (page 80). Following the establishment of Kew and other botanical gardens in the 18th and early 19th centuries, public collections of orchids gradually became available to be seen by the wider public, but private collections remained the preserve of the wealthy.

In the second half of the 20th century, mass propagation techniques were developed for some tropical orchids, which led to a major industry producing orchids for a wider community, often for little more than a bunch of cut flowers. Many people now buy orchids in garden centres and supermarkets to be grown on their windowsills. However, plants commonly available for purchase represent only a tiny proportion of the full diversity of orchids – most of the plants we can buy are commercially produced hybrids in a few genera, for example *Dendrobium* and *Phalaenopsis* (pages 55 and 87), selected for their suitability for cultivation.

In addition to their importance in horticulture, orchids are also widely used in foods and traditional medicines. The most widely used orchid product is the flavouring vanilla (derived from the fermented pods of *Vanilla* species), and this is used in chocolate, ice cream, cakes and many other foods and drinks. It is also used in scents. Less widely known food stuffs derived from orchids include *salep*, a starchy substance extracted from tubers of Mediterranean orchids including *Orchis mascula* (page 40) and *chikanda* made from tubers of some African species. In Asia, many orchids, including species of *Cremastra*, *Dendrobium* and *Gastrodia*, are used as traditional medicines. Natural vanilla is mostly produced from cultivated plants, but the other orchids used in food and some of the medicinal orchids are often still collected from the wild, and such harvest can represent a threat to the continued survival of the species involved.

The threat of over-collection for food, medicine and horticulture, combined with habitat destruction and more recently climate change, has led scientists in botanic gardens (including Kew) to expand their orchid studies to conservation. This involves a combination of *in situ* and *ex situ* methods and, in some cases, has resulted in the successful reintroduction of species that were on the verge of extinction. Despite these activities, assessments of threats to orchids using Red Listing and other techniques led by members of the Orchid Specialist Group of the International Union for the

Conservation of Nature, indicate that orchids are among the most threatened groups of plants. In extreme cases, nearly all species are threatened with extinction in the wild, a notable example being the tropical Asian lady's slipper orchids, *Paphiopedilum* (page 20).

Orchids are a fascinating group of plants, but they can also be seen as the plant equivalent of the 'canary in the coal mine' – due to their complex interactions with fungi and pollinators, many may be among the first casualties of declines in ecosystem health. If these orchids are to survive and continue to be enjoyed and used by future generations, we must continue to develop innovative conservation methodologies and make people aware of the importance of the impact we are having on these plants. Persuading people that plants from cultivation are more desirable than wild-collected plants will have a major role in conserving species that can be propagated under *ex situ* conditions, but for the continued existence of many orchids and their complex interactions in nature, preservation of their habits will be critical.

Michael F. Fay
Senior Research Leader, Conservation Genetics and Molecular Ecology
Royal Botanic Gardens, Kew
Co-Chair, Orchid Specialist Group, Species Survival Commission, IUCN

Apostasia odorata

grass orchid
from Nathaniel Wallich
Plantae Asiaticae Rariores, 1830–32

Apostasia odorata has a wide distribution in
East and Southeast Asia, and it is one of about
16 species in the two genera that make up the
smallest orchid subfamily, Apostasioideae. In
these species, unlike other orchids, the stamens
and style are not fully fused (a seemingly
primitive condition), but they are advanced
in other aspects.

Cleistesiopsis divaricata

rosebud orchid, small spreading pogonia
from Mark Catesby *Natural History of Carolina,
Florida, and the Bahama Islands*, 1754

———————

The rosebud orchid is a North American
species, found in wet meadows and pine-
dominated savannas on the coastal plain
from New Jersey to Florida and across the
Gulf Coast to Texas. It is a distant relative
of commercial *Vanilla*.

Cypripedium calceolus

lady's slipper orchid
from Rembert Dodoens *Stirpium Historiae
Pemptades Sex, Sive Libri* XXX, 1616

The lady's slipper orchid has one of the widest
distributions of any orchid (England to East
Asia), but it is on the verge of extinction in many
countries. In England, it now occurs at more
than a dozen sites, following a long-running
conservation project involving Kew and
other organisations.

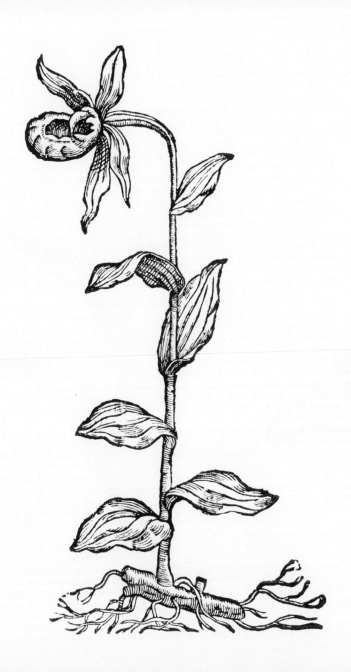

Cypripedium species

slipper orchid
by Georg Dionysius Ehret from
Kew Collection, 1756

———————

This illustration, originally labelled
Cypripedium calceolus, almost certainly
represents a North American member of the
C. pubescens/C. parviflorum group, as implied
by the presence of a North American black
swallowtail butterfly. The North American
plants have been shown to be distinct from
Eurasian *C. calceolus* by researchers
at Kew and elsewhere.

Paphiopedilum hookerae

slipper orchid
by Walter Hood Fitch from
Curtis's Botanical Magazine, 1863

Only found below sandstone cliffs on Borneo,
Hooker's slipper orchid was named for Maria
Hooker, the wife of Sir William Hooker (Kew's
first official Director) by Heinrich Reichenbach,
in recognition of her activities in supporting
her husband's work.

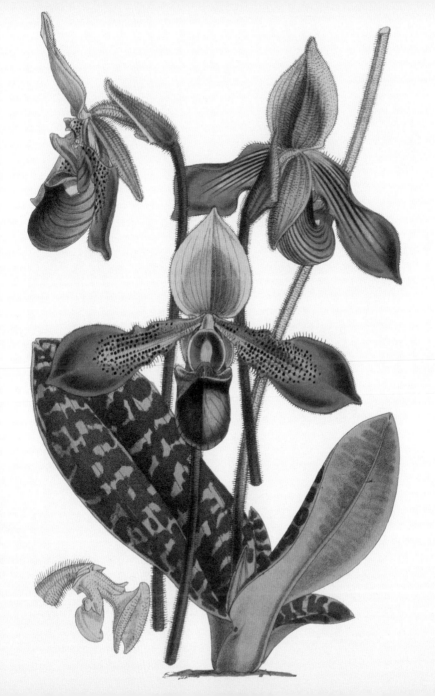

Phragmipedium warszewiczianum

slipper orchid
from R. Warner and B. S. Williams
The Orchid Album, 1889

Józef Warszewicz was a Polish botanist who
collected plants in Central and South America
for private collections and botanical gardens.
He was particularly interested in orchids, and
Reichenbach named this slipper orchid in his
honour. It is found on rocky outcrops and lava
flows in the central Andes.

Chloraea multiflora

gavilú

from Claudio Gay *Atlas de la Historia
Física y Politica de Chile*, 1854

One of about 50 species of *Chloraea*, this
terrestrial orchid occurs in southern and
central Chile. It is a distant relative of the
lady's tresses orchids. Its common name is
gavilú, but this name was used in its Latin
form (*Gavilea*) for a related group of
southern South American orchids.

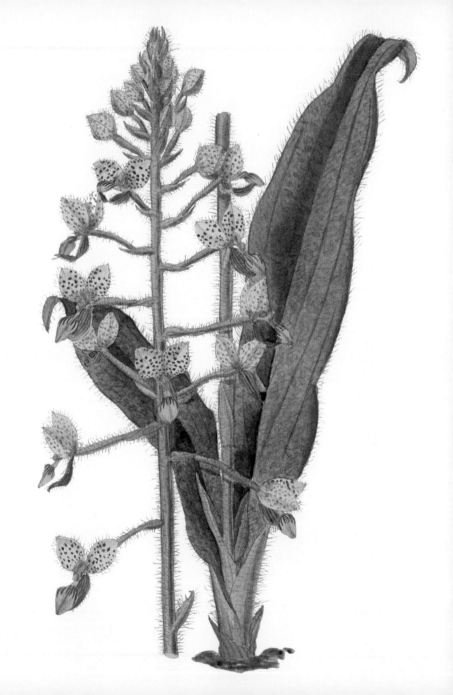

Ponthieva maculata

spotted ponthieva
by Matilda Smith from
Curtis's Botanical Magazine, 1882

Ponthieva maculata, native to Costa Rica,
Venezuela, Colombia and Ecuador, is one of
about 60 species in the genus. *Ponthieva* was
named after Henry de Ponthieu, a merchant
who sent plants from the West Indies to Sir
Joseph Banks (often referred to as Kew's first
unofficial director) in the 1770s. The epithet
maculata refers to the spotted petals.

Anoectochilus sp.

marbled jewel orchid, filigree orchid
from Prince of Wales Island (Penang)
Collection, Kew Collection, c.1800

With the common name of marbled jewel
orchids or filigree orchids in reference to the
leaves, *Anoectochilus* is a genus of about 50
species. Widespread in South and Southeast
Asia, some species occur as far afield as the
Himalayas, Australia and Hawai'i. They
prefer moist, shady places.

Corybas pictus

helmet orchid, painted corybas
from Ludwig von Blume *Collection des
Orchidées les Plus Remarquables de l'Archipel
Indien et du Japon*, 1858

With about 120 species, *Corybas* occurs in Asia,
Australasia and some Pacific islands. They are
known as helmet orchids due to the shape of
the flowers, and each plant has a single heart-
shaped leaf. *Corybas pictus* occurs in Malaysia
and Indonesia, growing in mossy vegetation in
forests from the foothills to 2000 m (6,562 ft).

Caladenia ixioides & C. gemmata

china orchids
from R. D. Fitzgerald
Australian Orchids, 1875–82

Commonly known as the white and blue china
orchids, respectively, these species are endemic
to Western Australia. They were previously
treated as members of *Cyanicula*. *Caladenia
ixioides* mostly occurs in woodlands near Perth.
Caladenia gemmata sometimes appear in large
numbers after fires.

Thelymitra longifolia

common sun orchid, maikuku
by George Bond from Kew Collection, 1823

The common sun orchid (*maikuku* in Maori) is
endemic to New Zealand, being found on both
main and some offshore islands. With a single
grass-like leaf and up to five white flowers, it
often grows in open areas in scrub, but can
also be found in forests.

Disa uniflora

Pride of Table Mountain
by Marianne North
from Marianne North Collection, Kew, 1882

———————

This South African orchid (the Pride of Table
Mountain), is restricted to mountains in the
southern Western Cape, where it grows near
waterfalls and streams. With showy scarlet
flowers, it is used as the logo for the Mountain
Club of South Africa. *Disa uniflora* was crossed
with *D. tripetaloides* at Kew in 1893, giving rise
to the popular hybrid *D.* ×*kewensis*. The blue
orchids in the painting are another *Disa*
species, possibly *D. graminifolia*.

Ophrys speculum

mirror orchid
by Walter Hood Fitch from
Curtis's Botanical Magazine, 1870

The mirror orchid is widespread in the
Mediterranean region. Like other *Ophrys*
species, it is pollinated by male insects (in this
case wasps) that try to mate with the flower,
mistaking it for a female. The "mirror" of the
orchid mimics the wings of the female, and
the floral scent mimics the sexual scent
of the female.

N⁰ 15

Orchis mascula

early purple orchid
by Frances Bauer from
Kew Collection, 18th century

———————————

The early purple orchid has a wide distribution
in Europe, North Africa and the Middle East. It
produces one rounded tuber each season, the
new one 'waxing' as the one from the previous
year 'wanes'. *Orchis* is the Greek word for
testicle (which the tuber resembles), and this
and related species were used as aphrodisiacs.

Sobralia macrantha

large-flowered sobralia
from Louis Van Houtte *Flore des Serres et des Jardins de l'Europe*, 1852

This species, occurring from Mexico to Costa Rica, produces large, fragrant (but short-lived) flowers. Along with many other orchids, *Sobralia macrantha* was named by John Lindley, Professor of Botany at the University of London. Lindley also played an important role in preventing the demise of Kew in the early days of Queen Victoria's reign.

Nervilia plicata

shield orchid

by unknown Indian artist commissioned by
William Roxburgh, Kew Collection, c.1800

One of about 65 species in the genus,
Nervilia plicata has a wide distribution in Asia,
and it also occurs in Queensland, Australia.
It produces a single leaf each year that is hairy
on each surface. After this leaf has withered,
the plant flowers. *Nervilia* refers to the
prominent veins/nerves on the leaf,
and *plicata* means folded.

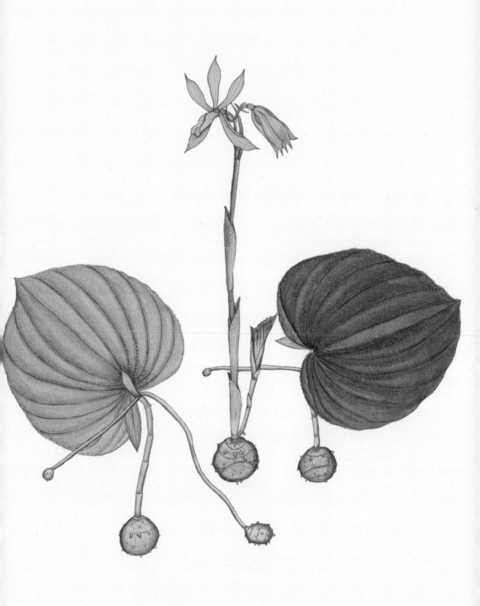

Epipogium aphyllum

ghost orchid
from Anton Kerner von Marilaun
Pflanzenleben, 1913–16

The ghost orchid *Epipogium aphyllum* is a
widespread Eurasian species. It produces no
leaves or chlorophyll and is fully dependent
on fungi for its food. It is notoriously difficult
to locate due to its lack of colour and sporadic
appearances. In England, it was rediscovered
in 2009 after being thought to be extinct
since the 1980s.

Arundina graminifolia

bamboo orchid
probably by unknown Chinese artist from
the General John Eyre Collection,
Kew Collection, mid-1800's.

———————

This reed-like plant, sometimes called the
bamboo orchid, is the only species in the genus.
With a wide native range in tropical Asia, it is
a popular garden plant in the tropics. It is one
of the few truly weedy orchids, having escaped
widely from cultivation.

Coelogyne asperata

rough-lipped coelogyne
by Marian Ellis Rowan commissioned by
W. H. Flavelle of Sydney for use on ceramics,
Kew Collection, early 20th century

———————

Coelogyne asperata is a widespread species
found on many islands in Southeast Asia to
Papua New Guinea and the Solomon Islands.
The fragrant flowers are reputed to smell
of liquorice. On the basis of DNA studies,
Coelogyne was recently expanded to include
some other genera and now comprises
more than 500 species.

Pleione hookeriana

Hooker's pleione
by Florence H. Woolward from
Curtis's Botanical Magazine, 1878

Originally described by Lindley as a species
of *Coelogyne*, *Pleione hookeriana* was named
from material collected by Joseph Dalton
Hooker (later the second Director of Kew) in
Darjeeling. It occurs from north-eastern India
to southern China, growing among bamboo
and *Rhododendron* at elevations up to
more than 4,200 m (13,780 ft).

Dendrobium densiflorum

pineapple orchid
by unknown Indian artist commissioned by
Nathaniel Wallich, Kew Collection, 1828

Dendrobium is one of the largest genera of
orchids, with about 1,200 species occurring
from India to New Zealand. The greatest
diversity is found in New Guinea. *Dendrobium
densiflorum* is an Asian species described in
1830 by John Lindley.

Cymbidium lancifolium

lance-leaved cymbidium
from William Jackson Hooker's Chinese Plants
Collection, Kew Collection, 1790–1850

This distinctive species was described by
William Hooker (later to become Kew's
first official Director), in 1823 from material
collected in Nepal by Nathaniel Wallich, who
was appointed Superintendent of the Calcutta
Botanic Garden on the recommendation of
Joseph Banks. The genus *Cymbidium* was
monographed by David Du Puy and
Phillip Cribb at Kew in the 1980s.

Eulophia guineensis

Guinea eulophia, broad-leaved ground orchid
by J. Curtis from
Curtis's Botanical Magazine, 1824

Eulophia is a genus of terrestrial orchids with more than 200 (many African) species. *Eulophia guineensis*, found across tropical Africa and in the Arabian Peninsula and Cape Verde islands, is the type species of the genus. It can reach a metre in height and occurs in woods and scrublands at elevations up to 2,000 m (6,562 ft).

Catasetum saccatum

sack-shaped catasetum
by Sarah Anne Drake, Kew Collection, 1837–41

Catasetum saccatum is native to Brazil, Guyana and Peru. *Catasetum* was one of the groups of orchids that particularly fascinated Charles Darwin, due to plants having separate male and female flowers (an unusual condition among orchids). He reported the explosive force with which pollen masses of *C. saccatum* are ejected when the male flowers are visited by bees.

Peristeria elata

Holy Ghost orchid, dove orchid
from *Lindenia: Iconographie des Orchidées*, 1899

Occurring from Costa Rica to Peru, the Holy
Ghost or dove orchid, *Peristeria elata*, is the
national flower of Panama. *Peristeria* comes
from the Greek word for a dove, in reference to
the shape of the lip and column of the flower.
Threatened by over-collection, it receives a high
level of protection under the Convention on
International Trade in Endangered Species.

Maxillaria acutifolia

rufous tiger orchid
by Hannah Im Thurn (née Lorimer) in Guyana,
Kew Collection, 19th century

Maxillaria acutifolia occurs in wet forests at
190–2,300 m (523–7,546 ft) above sea level
in Mexico, Central America, northern South
America and some Caribbean islands. Its
yellow-orange flowers have a pleasant scent.
Maxillaria refers to the shape of the column in
some species being like a jawbone [= maxilla],
and *acutifolia* refers to the pointed leaves.

Miltonia clowesii

Clowes' miltonia
by Edith Holland Norton,
Kew Collection, 1880–82

Miltonia clowesii is only known from south-eastern Brazil. Originally described by Lindley as a species of *Odontoglossum*, it has a chequered taxonomic history, having been included in three other genera. This reflects the difficulty in defining genera in the larger group including *Miltonia*.

Coryanthes macrantha

bucket orchid
by Sarah Anne Drake from
Edward's Botanical Register, 1836

Coryanthes macrantha, the bucket orchid,
produces large flowers (up to 15 cm / 6 inches),
and is native to Trinidad and tropical South
America. It has a peculiar pollination
mechanism, in which male euglossine bees
collect the floral scent (to attract females), in the
process falling into the 'bucket'. To escape, they
must crawl through a narrow passage, picking up
the pollen in the process.

Pescatoria lehmannii

Lehmann's pescatoria
from R. Warner and B. S. Williams
The Orchid Album, 1883

The candy-striped, waxy flowers of *Pescatoria lehmannii* can be up to 8.5 cm (3½ inches) across. They are strongly scented and have a very hairy lip. The species is found in cloud forests in the coastal mountains of southern Colombia and Ecuador.

Cattleya cernua

nodding sophronitis
from Alfred Cogniaux *Dictionnaire Iconographique des Orchidées*, 1896–1907

Cattleya (previously *Sophronitis*) *cernua* occurs from southern Brazil to northern Argentina. Like many Neotropical orange- or red-flowered species, it is pollinated by hummingbirds. *Sophronitis* was merged with *Cattleya* following doctoral research conducted at Kew by Cássio van den Berg. Lindley named *Cattleya* after William Cattley who flowered *C. labiata* (the corsage orchid) in the UK in the early 19th century.

Epidendrum paniculatum

paniculate epidendrum
from Kew Collection, 19th century

Epidendrum paniculatum in the narrow sense
is only found in wet forests between 1,700 and
2,800 m (5,577–9,186 ft) in Peru. Several other
species were previously included in this one,
resulting in a much wider distribution. It is
adapted to pollination by butterflies. *Epidendrum*,
with more than 1,500 species in the Neotropics, is
one of the largest genera of orchids.

Dracula chestertonii

frog's skin
by Matilda Smith from
Curtis's Botanical Magazine, 1888

Carlyle Luer described the genus *Dracula*
in 1978 for a group of species previously
included in *Masdevallia*. *Dracula* means 'little
dragon' and refers to the strange flowers,
not the infamous Count! From Colombia, *D.
chestertonii* was named after its collector
James Henry Chesterton; he collected many
orchids in South America for the famous
Veitch nurseries in Chelsea.

Masdevallia rosea

pink masdevallia
from *La Belgique Horticole: Annales de Botanique et d'horticulture*, 1882

———————

Masdevallia rosea (the pink masdevallia) is native to the Andes of Colombia and Ecuador, where it grows in cloud forests at 2,400 – 3,400 m (7,874 – 11,155 ft). Its brightly coloured flowers are adapted to pollination by hummingbirds. Each inflorescence produces only a single flower.

Calanthe tankervilleae

**greater swamp orchid, swamp lily,
nun's orchid, veiled orchid**
probably by unknown Chinese artist from
the General John Eyre Collection, Kew
Collection, mid-1800's

Native to tropical Asia, Australasia and
some Pacific islands, *Calanthe* (previously
Phaius) *tankervilleae* is also naturalised in the
Neotropics and Hawai'i. It was the first tropical
orchid to be flowered in the UK, and it was
named after Lady Emma Tankervilļe, in whose
glasshouse this achievement occurred.

Calanthe ×veitchii

Veitch's Christmas orchid
by John Day from John Day Scrapbooks,
Kew Collection, 1883

A popular horticultural hybrid, *Calanthe
×veitchii* was produced in the Veitch Nurseries
in Chelsea and was registered with the Royal
Horticultural Society in 1860. The parents
were *C. rosea* (female) and *C. vestita* (male), the
former native to Myanmar and Thailand and
the latter occurring from Assam to New Guinea.

Vanda falcata

wind orchid
from William Jackson Hooker's Chinese Plants
Collection, Kew Collection, 1790–1850

Vanda (previously *Neofinetia*) *falcata* is native
to China, Korea and Japan. Many cultivars
exist, varying in vegetative and floral characters.
Once the preserve of the rich in Japan, where
they are known as *fūkiran* ('orchid of rich and
noble people'), they increased in popularity in
the early 20th century. They are often grown in
specially designed pots.

Phalaenopsis amabilis

moth orchid
by Marianne North from
Marianne North Collection, Kew, 1870

The lovely phalaenopsis or moth orchid,
Phalaenopsis amabilis, has a wide distribution
including Southeast Asia, Papua New Guinea
and northern Australia. The Australian
subspecies is threatened by over-collection. The
species is a progenitor of nearly all the hybrid
moth orchids that now form the basis of a huge
international pot plant industry.

Renanthera coccinea

probably by unknown Chinese artist from
the General John Eyre Collection,
Kew Collection, mid-1800's

Found in China, Laos, Myanmar, Thailand
and Vietnam, *Renanthera coccinea* is the type
species of the genus. It was described by the
Portuguese Jesuit missionary João de Loureiro.
He spent 40 years in Cochinchina (now part
of Vietnam) in the 18th century, and his
Flora Cochinchinensis contains descriptions
of this and many other Vietnamese plants.

Angraecum sesquipedale

comet orchid

from R. Warner and B. S. Williams
The Orchid Album, 1887

Angraecum sesquipedale, endemic to lowland
forests in Madagascar, is known as the comet
orchid due to the length of its spur. Charles
Darwin famously predicted the existence of a
hawkmoth with a 'wonderfully long proboscis'
capable of reaching the nectar at the end
of the spur.

ILLUSTRATION SOURCES

Books and Journals

Blume, Ludwig von. (1858). *Collection des Orchidées les Plus Remarquables de l'Archipel Indien et du Japon*. C. G. Sulpke, Amsterdam.

Catesby, Mark. (1754). *Natural History of Carolina, Florida, and the Bahama Islands*. C. Marsh, London.

Cogniaux, Alfred. (1896–1907). *Dictionnaire Iconographique des Orchidées*. F. Havermans, Brussels.

Dodoens, Rembert. (1616). *Stirpium Historiae Pemptades Sex, Sive Libri XXX*. Ex Officina Plantiniana, apud Balthasarem et Ioannem Moretos, Antwerp.

Fitzgerald, R. D. (1875–82). *Australian Orchids*. Thomas Richards, Sydney.

Gay, Claudio. (1854). *Atlas de la Historia Física y Politica de Chile*. E. Thunot y Ca, Paris.

Hooker, J. D. (1863). *Curtis's Botanical Magazine*. Volume 89, t. 5362.

Hooker, J. D. (1870). *Curtis's Botanical Magazine*. Volume 96, t. 5844.

Hooker, J. D. (1878). *Curtis's Botanical Magazine*. Volume 104, t. 5388.

Hooker, J. D. (1882). *Curtis's Botanical Magazine*. Volume 108, t. 6637.

Hooker, J. D. (1888). *Curtis's Botanical Magazine*. Volume 114, t. 6977.

Kerner von Marilaun, Anton. (1913–16). *Pflanzenleben*. Bibliographisches Institut, Leipzig.

Linden, L. (1899). *Lindenia: Iconographie des Orchidées*. Lucien Linden, Brussels.

Lindley, J. (1836) *Edwards's Botanical Register*. Volume 22, t. 1841.

Morren, Edouard. (1882). *La Belgique Horticole: Annales de Botanique et d'Horticulture*. La Direction Générale, Liége.

Regel, E. (1856). *Gartenflora*. F. Enke, Erlangen.

Sims, J. (1813). *Curtis's Botanical Magazine*. Volume 37, t. 1512.

Sims, J. (1824). *Curtis's Botanical Magazine*. Volume 51, t. 2467.

Van Houtte, Louis. (1852). *Flore des Serres et des Jardins de l'Europe*. Louis van Houtte, Gand.

Wallich, Nathaniel. (1830–32). *Plantae Asiaticae Rariores, or, Descriptions and Figures of a Select Number of Unpublished East Indian Plants*. Treuttel and Würtz, London, Paris, Strasbourg.

Warner, R. and Williams, B. S. (1883, 1887 and 1889). *The Orchid Album*. B. S. Williams, London.

Art Collections

John Day (1824–1888). An English orchid-grower and collector, who produced some 4,000 illustrations of orchids in 53 scrapbooks, dating from 1863 until his death. The scrapbooks were donated to Kew in 1902 by his sister, Emma Wolstenholme.

Sarah Anne Drake (1803–1857). Commissioned by John Lindley, botanist and secretary of the Horticultural Society of London, to draw numerous plant specimens. His *Sertum Orchidaceum* (1837–1841) boasts some of her most stunning illustrations. Kew holds most of the original drawings prepared by Drake for this work.

General John Eyre (1791–1865). Collection comprising 190 watercolours by unknown Chinese artists commissioned by Eyre during his time stationed in Hong Kong as the Commander of the British Forces in China, 1847–51.

Marianne North (1830–90). Over 800 oils on paper, showing plants in their natural settings were painted by North, who recorded the world's flora during travels from 1871 to 1885, with visits to 16 countries in 5 continents. The main collection is on display in the Marianne North Gallery at Kew Gardens, bequeathed by North and built according to her instructions and first opened in 1882.

William Roxburgh (1751–1815). 'Icones': one of two sets of comprising 2,500 drawings by unknown Indian artists, made 1776–1813, working on the Coromandel Coast and at the Calcutta Botanic Garden. A duplicate set is held by the Central National Herbarium, Botanical Survey of India, in Kolkata's Acharya Jagadish Chandra Bose Indian Botanic Garden.

Nathaniel Wallich (1786–1854). About 1,000 drawings by unknown Indian artists, intimately connected to the Herbarium of the East India Company, also known as the Wallich Herbarium, held at Kew, comprising dried plant specimens obtained by Wallich on his travels in the Indian subcontinent.

FURTHER READING

Chase, Mark W, Christenhusz, Maarten J. M., Mirenda, Tom. (2017). *The Book Of Orchids: A Life-Size Guide to Six Hundred Species From Around The World.* Ivy Press, Brighton and University of Chicago Press, Chicago.

Christenhusz, Maarten J. M. Fay, Michael F. and Chase, M. W. (2017). *Plants of the World. An Illustrated Encyclopedia of Vascular Plant Families.* Royal Botanic Gardens, Kew and University of Chicago Press, Chicago.

Endersby, Jim. (2016). *Orchid: A Cultural History.* Royal Botanic Gardens, Kew and University of Chicago Press, Chicago.

Fay, Michael F. (2015). British and Irish Orchids in a Changing World. *Curtis's Botanical Magazine*. Volume 32.

Gardiner, Lauren and Cribb, Phillip. (2018). *The Orchid*. Book with 40 art prints. Welbeck Publishing, London in association with the Royal Botanic Gardens, Kew.

Kanellos, Manos and White, Peter. (2020). *Growing Orchids at Home*. Royal Botanic Gardens, Kew.

Kirby, Stephen, Doi, Toshikazu and Otsuka, Toru. (2018). *Rankafu Orchid Print Album:Masterpieces of Japanese Woodblock Prints of Orchids*. Royal Botanic Gardens, Kew.

Kühn, Rolf, Pedersen, Henrik Ærenlund, Cribb, Phillip. (2019). *Field Guide to the Orchids of Europe and Mediterranean*, Royal Botanic Gardens, Kew.

North, Marianne and Mills, Christopher. (2018). *Marianne North: The Kew Collection*. Royal Botanic Gardens, Kew.

Rix, M. (2021). *Indian Botanical Art: An Illustrated History*. Royal Botanic Gardens, Kew and Roli Books, New Delhi.

Seaton, Philip. (2020). *The Kew Gardener's Guide to Growing Orchids*. Frances Lincoln in association with the Royal Botanic Gardens, Kew.

Willis, Kathy and Fry, Carolyn. (2014). *Plants from Roots to Riches*. John Murray, London in association with the Royal Botanic Gardens, Kew.

Online

www.biodiversitylibrary.org – biodiversity and natural history literature including many rare books.

www.kew.org – information on Kew's science, collections and visitor programme.

www.plantsoftheworldonline.org – authoritative information on the world's flora from the botanical literature.

ACKNOWLEDGEMENTS

Michael F. Fay was introduced to the wonderful world of orchids by the late Amy Morris in the 1960s, starting a lifelong passion for these plants. His husband, Mark Chase, is thanked for all the discussions about orchids over the last 30 years. Hassan Rankou led the work on conservation assessments, notably for the slipper orchids.

Kew Publishing would like to thank the following for their help: in Kew's Library and Archives, Fiona Ainsworth, Julia Buckley, Anne Marshall, Cecily Nowell-Smith, Lynn Parker; for digitisation work, Paul Little.

INDEX

First published in 2022
Royal Botanic Gardens, Kew,
Richmond, Surrey, TW9 3AB, UK
www.kew.org

ISBN 978 1 84246 771 8

Distributed on behalf of the Royal Botanic Gardens, Kew in North America by the University of Chicago Press, 1427 East 60th St, Chicago, IL 60637, USA.

British Library Cataloguing in Publication Data
A catalogue record for this book is available from the British Library

Design: Ocky Murray
Page layout and image work: Christine Beard
Production Manager: Jo Pillai
Copy-editing: Ruth Linklater

Printed and bound in Italy by Printer Trento srl.

Front cover image: *Vanda tricolor* by Pieter de Pannemaeker from J. J. Linden *Lindenia: Iconographie des Orchidées*, 1886.

Endpapers: *Vanda coerulea*, blue orchid from the John Day Scrapbooks, Kew Collection, 1880.

p2: *Satyrium carneum* by S. T. Edwards from *Curtis's Botanical Magazine*, 1813.

p4: Orchid species including *Cypripedium*, *Orchis*, *Anacamptis* and *Ophrys* from E. Regel *Gartenflora*, 1856.

p10–11: *Dendrobium album* by Margaret Read-Brown, Kew Collection, 1862.

For information or to purchase all Kew titles please visit shop.kew.org/kewbooksonline or email publishing@kew.org

Kew's mission is to understand and protect plants and fungi, for the wellbeing of people and the future of all life on Earth.

Kew receives approximately one third of its funding from Government through the Department for Environment, Food and Rural Affairs (Defra). All other funding needed to support Kew's vital work comes from members, foundations, donors and commercial activities, including book sales.

Publishers note about names
The scientific names of the plants featured in this book are current, Kew accepted names at the time of going to press. They may differ from those used in original-source publications. The common names given are those most often used in the English language, or sometimes vernacular names used for the plants in their native countries.